FROM THE HILL : THE STORY OF

LOWELL
OBSERVATORY

FROM THE HILL : THE STORY OF

LOWELL
OBSERVATORY

SECOND EDITION

Lowell Observatory
Flagstaff, Arizona

Rose Houk

Designed by Ormsby and Thickstun
Production by Northland Graphics, Flagstaff, Arizona

Acknowledgments

The entire staff of Lowell Observatory deserves thanks for their willingness to help in the research for this book. Trustee William Lowell Putnam enthusiastically embraced the idea of the book. Observatory director Robert Millis was exceptionally supportive throughout the process, as were all the astronomers who patiently answered questions, gave tours, and made apparent the wish that their work be accessible to the general public. Librarian Antoinette Beiser was always willing to assist, as were all the support staff. Cynthia Webster ably oversaw editorial coordination and final production.

Contents

Lowell Observatory invites the public to visit Mars Hill on the west side of Flagstaff. The Steele Visitor Center is open daily in the summer with limited hours in winter. Exhibits and tours are available as part of daytime and nighttime programs. Night programs feature viewings on the Clark twenty-four-inch and McAllister sixteen-inch telescopes, weather permitting. Group tours are also possible with advance reservations. Separate day and nighttime admission fee.

For information about public programs call (520) 774-2096. For group reservations and information about special programs or the portable planetarium, call (520) 774-3358. Website: http://www.lowell.edu/ Address: 1400 West Mars Hill Road, Flagstaff, Arizona, 86001

Lowell Observatory is a private, nonprofit research facility whose public program is supported primarily by admission fees and donations.

Steele Visitor Center, Mars Hill

Percival Lowell, founder of Lowell Observatory

Almost a century ago, a telegram

was fired off from the cosmopolitan eastern city of Boston to a backwoods lumber town in the Arizona territory: "Site probably Flagstaff. Prospect for best seeing. Report climate... compared with Tucson."

The telegram was sent on April 10, 1894, by Percival Lowell to Andrew Ellicott Douglass. A young Harvard-trained astronomer and son of an Episcopalian minister, A.E. Douglass was a man with a mission. He had been sent by Mr. Lowell to the western part of the country to find the best location for an astronomical observatory. All of Douglass's work was to be completed in time for a favorable view of Mars that was to occur one short month later. Douglass had already tested Tombstone, Tucson, Tempe, and Prescott and found them wanting in one of the most critical criteria of astronomical observing—the illusive quality called "seeing."

Flagstaff fulfilled both men's fondest wishes that spring. A.E. Douglass followed Percival Lowell's instructions; he set about building an observatory on a hill on the outskirts of the small town of Flagstaff, population a thousand souls, tucked in the ponderosa pine forest of northern Arizona.

Lowell Observatory has since become a household word in the world of astronomy. From its home on Mars Hill have been conducted thorough searches for life on another planet; Pluto, our solar system's ninth planet, was discovered there; and pioneering observations that contributed to current concepts of the universe were also made at Lowell.

But Lowell Observatory is more than a quaint example of nineteenth-century science. Today its multitude of telescopes is still pointed at the night sky, as the observatory's staff of scientists continues research on basic questions about the solar system, and the stars and galaxies far beyond. In addition, Lowell Observatory serves as a training ground where university students can gain hands-on experience in astronomy.

Percival Lowell dedicated the last twenty-two years of his life to work at his observatory. He was there when he died in 1916 and is buried in a mausoleum that looks out toward the towering San Francisco Peaks. His philosophy as an astronomer is etched in the gray New England granite of his tomb: "To see into the beyond requires purity... and the securing it makes him perforce a hermit from his kind.... He must abandon cities and forego plains... only in places raised above and aloof from men can he profitably pursue his search."

Percival Lowell made a fine choice in his selection of a place — high above the cities and the plains — where he would carry out the final passion of his life and where his endowment would allow others to do likewise.

A.E. Douglass in the dome with the Clark telescope

Flagstaff It Is

EARLY TELESCOPE ON MARS HILL

A.E. Douglass started his search for the site of Percival Lowell's observatory in March 1894 in southern Arizona. He first tried Tombstone, then moved on to Tucson, where he found not only bad weather but troublesome mosquitoes as well. In daily dispatches to his employer in Boston, he reported vital statistics of each place he sampled — rainfall, altitudes, winds, and the "seeing."

"Seeing" is an intrinsic quality of the air, concerned not so much with clarity as with steadiness. Heavenly objects, when seen through the turbulence of Earth's atmosphere, are given to wobbling in the eyepiece of a telescope. For astronomers, the more this pesky problem of atmospheric turbulence can be avoided the better, hence the emphasis on good seeing.

Percival Lowell, acutely aware of the importance of seeing, bore it in mind constantly during the dogged search for an observatory site. The yardstick he and Douglass used to measure seeing was a scale devised by astronomer W.H. Pickering of the Harvard Observatory, who was also involved in the founding of Lowell Observatory. Like the scoring in a diving contest, Pickering's scale of seeing progressed from one to ten, with ten being the most desirable.

Finding nothing above a seven in the southern part of the state, Douglass headed north, aiming always for a ten. On the evening of April 4, with Lowell's own portable six-inch telescope, he launched a two-week series of observations in several places around Flagstaff. Douglass sent reports almost daily to Lowell, including details about the town. Although Flagstaff lacked a river of any size, it did have a "good hotel for this part of the U.S. Ladies can stay there." More important, he found a site that offered especially good seeing. As time became more critical, Lowell wired Douglass stating Flagstaff as the tentative site. Following another week of reports from Douglass, Lowell made up his mind and telegraphed: "Flagstaff it is." The observatory would be on site 11, situated on a low hill just west and 330 feet

The small town of Flagstaff, circa 1896 to 1900

above town. Lowell then admonished Douglass, as he would often in the ensuing months, to push on with work on the observatory "as fast as possible."

Douglass wasted no time. Ground was broken on April 23 for the foundation of the dome that would house the telescope, to arrive later from Boston. Twenty-four juniper posts were sunk into the basalt boulders that cap the mesa, and the posts were covered with thin boards. The top of the dome would revolve on wheels set into a hardwood track, so that one man could move it easily with a set of block and tackle. The shutter opening in the roof was eight feet wide, covered with a canvas curtain that happened to work well in

summer but behaved "mischievously" in winter.

Douglass was maintaining a hectic schedule, not only with construction but with astronomical observations as well. While the telescope pier was being installed, he was sending Lowell drawings and reports of his search for Gale's Comet. Douglass was also keeping close accounts of the money being spent, and finally had to ask Mr. Lowell to forward a thousand dollars to pay mounting expenses. His employer obliged, advising Douglass to take out two hundred dollars, a fourth of his salary.

Meanwhile, the fledgling town of Flagstaff — soon dubbed the Skylight City — guaranteed title to five acres of land for the observatory and pledged to build a wagon road to the top of the mesa.

May 28, 1894, was a monumental day in the history of Lowell Observatory. William Henry Pickering and Percival Lowell himself arrived in Flagstaff. Borrowed eighteen- and twelve-inch telescopes were in place in the new dome, and all was in readiness to observe a "favorable opposition" of Mars — when it is opposite the Sun and close to the Earth.

Observations of the planet began immediately and continued almost nightly for the next six months. But then things began to turn sour. Just as Lowell was leaving in December, Flagstaff's famous wintry weather set in with a vengeance. Lowell later sent condolences to Douglass as the weather, along with the observing, grew grim beyond words.

At times snowfall was so heavy that Douglass had to climb a ladder and enter the telescope dome through the shutter, and he could toboggan on his snowshoes down the hill to the boarding house where he stayed. Flagstaff handyman Stanley Sykes, who would become Lowell Observatory's instrument maker, related to Douglass his own scale of seeing, more appropriate to Flagstaff conditions: "10 is when you can see the moon, 5 is when you can still see the telescope, and 1 is when you can only feel the telescope but not see it."

By March Lowell said if such conditions continued he could "see little use in keeping up the observatory any longer." He was on the verge of returning the telescope, dismantling the dome, and relocating in Mexico. And that he did. By the middle of April 1895 Lowell Observatory officially closed — but only temporarily.

Percival Lowell continued an on-again, off-again search for an ideal observatory site. He looked all over the world — including not only Mexico, but the Sahara Desert and the Andes Mountains as well. But even

with its imperfections, Flagstaff eventually won out because overall it was judged better than any other place Lowell had tried.

In the summer of 1896 A.E. Douglass was back in Flagstaff, and later that year a fine new twenty-four-inch refractor telescope and a handsome forty-foot dome to house it in were returned from Mexico to be placed on Mars Hill. The telescope was built by the

The dome that houses the twenty-four-inch Clark telescope

famous firm of Alvan Clark and Sons, and the dome had been constructed by Godfrey Sykes, who worked with his brother Stanley in their "Makers and Menders" business in town. Lowell arrived on July 22, installed the lens in the telescope, and again began immediate observations of Mars.

His subsequent work on Mars brought Percival to Flagstaff often; it became his home away from home. From the busy streets of Boston he would return to Mars Hill and tend a garden that grew legendary pumpkins and squash. He took an intense interest in everything around him and spent most of his spare time exploring places around Flagstaff. He was an amateur botanist as well, dutifully noting the blooming times of wildflowers on Mars Hill and sending specimens of new plants to Dr. Charles S. Sargent at the Arnold Arboretum.

Percival Lowell's style of living reflected his personal wealth. In all his portraits, gazing steady-eyed into the camera, he presents an image of sartorial excellence — dressed in starched white shirt and three-piece tweed suit and jauntily holding a walking cane. Distinguished observatory guests who debarked from the Santa Fe Railroad in Flagstaff were fortunate recipients of Percival's generous hospitality. These people — paleobotanist Lester Frank Ward, newspaper magnate William Randolph Hearst, and legislator Henry Fountain Ashurst to name a few — always sat down at tables bedecked with fine linens and glistening crystal.

Despite his many diversions, Percival Lowell never strayed from his prime astronomical interest — Mars. The red planet was the reason he had founded Lowell Observatory, and before long he had formulated a comprehensive theory about it. That theory aroused vigorous controversy in the astronomical community and fired a public fever that at times appeared incurable.

Top: The Baronial Mansion
Bottom: Percival Lowell, far right, in his study with Wrexie Louise Leonard and Edward S. Morse

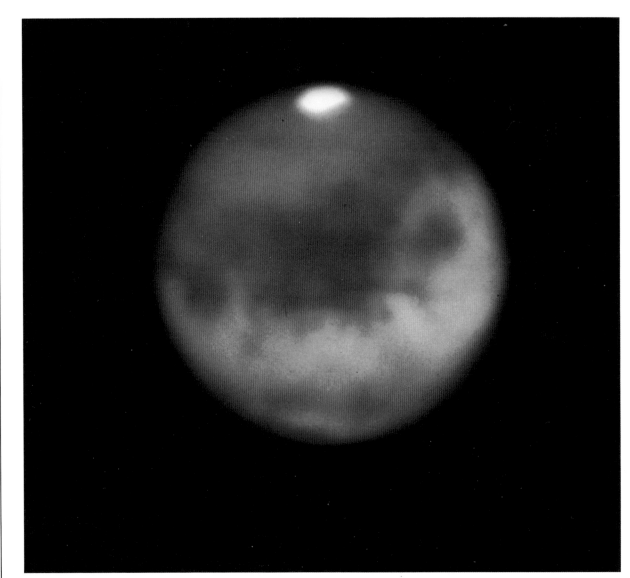

Photo of the planet Mars taken at Mauna Kea Observatory in 1988. Image on this and facing page printed with south up to correspond with historic drawings.

Lowell and Mars

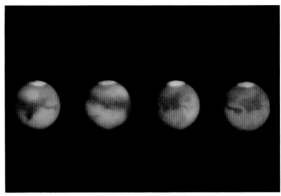

THE FACES OF MARS

Percival Lowell expressed no doubts about the probability of life on Mars. Entering the field of astronomy when he did, in the late nineteenth century, he inherited earlier views of Earth's near neighbor. Maps of Mars in those days showed ice caps, seas and lakes, and even continents and canals. With attributes so similar to Earth's, it did not require a great leap of faith to assume that life could exist there.

It was this curiosity about martian existence that spurred Lowell to found an observatory where he could pursue his major interest. Beginning in 1894 when Mars was close to the Earth, and for the next twenty-two years, he collected vast quantities of data, sketches, and photographs of the planet, and published volumes of scientific papers, popular articles, and books on his theory.

People have wondered why this wealthy man, at age thirty-nine, suddenly embarked upon a career in astronomy with little formal training in the discipline.

Lowell himself recalled that his earliest interest in astronomy and Mars dated well before 1894, to 1870 when he said he "used to look at Mars with as keen interest as now." As a young man he read astronomy books, had his own telescope, and gazed at the stars from the roof of his parents' house in Brookline, Massachusetts.

Lowell, whose family name is attached to the Massachusetts textile town, was born in 1855. His upbringing reflected his Bostonian parents' wealth and social stature — Percival attended preparatory schools in France, graduated Phi Beta Kappa from Harvard, learned about investments and business in his grandfather's cotton mill, and then traveled and lived in the Orient. Despite release from the everyday worry of earning a living, he was not one of the idle rich. His father had ingrained in his children the fundamental necessity that "every self-respecting man must work at something... of real significance."

In 1893, his fascination with the Orient sated, Percival Lowell returned to Boston. Lawrence, his younger brother and biographer, said that Percival "left no statement of why he gave up Japan for astronomy." Perhaps he had learned all he could there, or astronomy may have substituted as an intellectual interest. Interestingly, Percival had taken a telescope to Japan with him, the same six-inch that A.E. Douglass would use on his reconnaissance trip to Arizona.

When Lowell arrived at his new observatory in 1894, he was already fully aware of the work of Italian astronomer Giovanni Schiaparelli. In 1877 Schiaparelli had seen lines on Mars which he called *canali,* or "channels," which soon mistranslated into "canals." The implication was made explicit: such features could be the work of intelligent beings.

Lowell said of the network of canals that "A mind of no mean order would seem to have presided over the system we see — a mind certainly of considerably more comprehensiveness than that which presides over the various departments of our own public works." He noted that the lines were exceedingly uniform in size and pattern — they averaged thirty miles wide, were "fine and straight," and radiated from special points. To him they "certainly" represented an irrigation system, much needed on Mars because of its lack of water. The beings that masterminded this artificial network inhabited an arid planet, whose water came mostly in the form of dew or frost and seasonal meltings of the polar ice caps.

Would other physical conditions on Mars allow life to exist? Lowell's observations and calculations led him to conclude that Mars indeed had an atmosphere that could support life. That atmosphere was thin, cloudless, and calm, with a mean temperature of forty-eight degrees Fahrenheit, he said. The thin atmosphere was not a problem because Lowell did not think lungs were a prerequisite to intelligent life: "There is nothing in the world or beyond it," he wrote, "to prevent . . . a being with gills, for example, from being a most superior person."

Martians as little green men was not an image that Percival Lowell would have supported — at least not the part about *men.* Though he never wavered in his belief that life probably existed on Mars, Lowell never

Percival Lowell's early sketches of Mars, made in June 1894

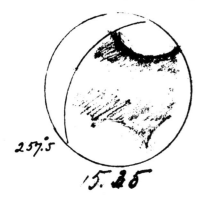 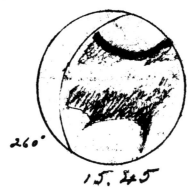 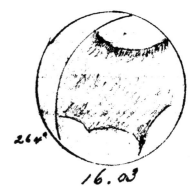

speculated what form that life took. To his chagrin, throughout the Mars furor the public and the press usually failed to acknowledge that important qualification.

Even as early as 1894 and 1895, after only a few months of observation, Percival Lowell began to publish his theory of Mars. The first full explanation appeared in a series in *Popular Astronomy,* followed by another series of articles in the *Atlantic Monthly.* He also wrote three major books on the subject: *Mars* in 1895, *Mars and Its Canals* in 1906, and *Mars as the Abode of Life* in 1908. Lowell was a superlative stylist and punster, and his adeptness at writing for a popular audience made his ideas all the more palatable to an eager public.

Reaction to his theory ran the gamut — the astronomical community ranged from vehement criticism through mild objection to some support. Many astronomers simply did not see the same markings on Mars that Lowell did; often, however, they were as critical of his popularization as they were of the theory itself. But Lowell insisted that scientists were obliged to present their work to the public in clear,

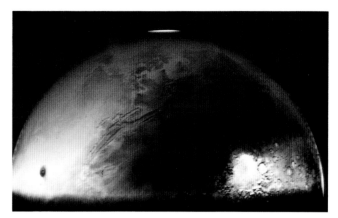

Viking spacecraft image of Mars

understandable terms. The press often responded with wry humor and occasional embellishments, while the public could not seem to get enough.

Percival Lowell's breakneck pace of observing and traveling and writing finally took its toll. In 1897 he suffered a breakdown from nervous exhaustion, forcing him into a lengthy period of convalescence. Not until 1901 was he well enough to return to his observatory and resume his work.

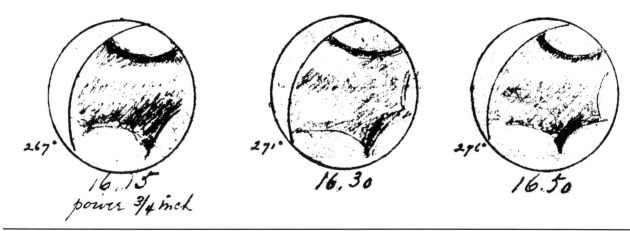

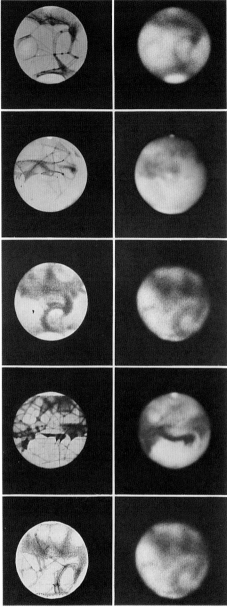

Drawings of Mars in left column compared to photographs of the planet in the right column, taken at approximately the same times

During Lowell's absence, some things had changed at the observatory. The eccentric Dr. Thomas Jefferson Jackson See for a time was put in charge in Flagstaff, and his personality quirks drove most of the assistants from the observatory. In 1901, Lowell and A.E. Douglass came to an unfortunate and cloudy parting of the ways as well. Douglass's star was rising, however, and he went on to become director of Steward Observatory in Tucson and to found the science of dendrochronology, or tree-ring dating.

Upon his recovery Lowell returned to Flagstaff and reentered the fray with characteristic vigor. In 1903 another opposition of Mars occurred. Lowell made productive observations and some new information resulted. In 1905 he revealed that Carl Otto Lampland, who had joined the observatory staff three years earlier, had taken the first photographs of Mars that seemed to show the controversial canals. Lowell was ready for another opposition of the planet in 1907, and in that year he sponsored an expedition to South America and spent eight months in Flagstaff observing Mars.

The year 1907 marked a turning point, however. As author William Hoyt observed, it was a turning point not only in the furor over Mars but also in Lowell's career. "Where before he had aggressively proclaimed his theories and confidently brushed aside the skepticism they stirred, he was now everywhere on the defensive," Hoyt wrote. The public had begun to lose interest in Mars, and Lowell found some difficulty getting his news published. As new revelations became more scarce, Lowell assumed a more philosophical stance, attributing attacks on his theory to people's inability to accept new ideas.

Not until the 1970s would Lowell's theory of Mars really be tested. The findings of the Mariner space missions, particularly Mariner 9 in 1971, showed a

V.M. Slipher, right, looks through Clark telescope, while C.O. Lampland records data

planet marked by vast craters and enormous volcanoes. Lowell was correct in his belief about the polar ice caps on Mars, but rather than consisting of water they appear to be mostly frozen carbon dioxide. The atmosphere is indeed thin, but it is not calm. Instead Mars is subject to great winds that stir up violent dust storms. Temperatures on Mars can vary wildly from minus 225 degrees to 63 degrees Fahrenheit. Percival's famous "canals" have never been documented by any of the recent space missions. Channels that appear to be ancient riverbeds have been mapped, but they are unrelated to the canals.

We possessed one strategic tool in our arsenal that Lowell did not have—the ability to land on Mars and obtain samples of actual soil from its surface. The samples gathered by the Viking lander in 1976 yielded no evidence of any organic molecules, dashing hopes that life as we know it exists on Mars.

Though Percival Lowell's theories about Mars have been proved largely wrong, nevertheless if we go back to the first decades of the twentieth century, we see him working productively on many fronts at his new observatory. Furthermore, Mars was not the only planet he studied: Mercury, Venus, and Saturn were also investigated, and Lowell quietly launched a search for a planet beyond Neptune that would lead to the discovery of Pluto. And during those decades one of astronomy's most significant discoveries was made by a man Lowell hired in 1901.

Andromeda Nebula

AN EXPANDING UNIVERSE

An Expanding Universe

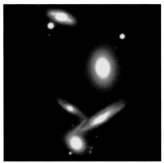

COMPACT QUINTET OF GALAXIES

Vesto Melvin Slipher was hired by Percival Lowell in 1901 as a "temporary" assistant. Fifty-three years later he retired as the observatory's second director.

Slipher, an Indiana farm boy and country school teacher, had impressed his professor Wilbur Cogshall. Cogshall, who had served on the Lowell Observatory staff, persuaded Percival Lowell to take the young Slipher aboard, but Lowell did so without great enthusiasm. "I shall not want another permanent assistant and take him only because I promised to do so," Lowell wrote Cogshall.

But V.M. Slipher soon distinguished himself as a great asset to Mr. Lowell's observatory. He was intelligent, patient, resourceful — and as cautious as his boss was impetuous. Slipher's first assignment was to learn how to use the observatory's new spectrograph — an instrument attached to a telescope that disperses incoming light into colors to produce a spectrum, much as water droplets do to make rainbows.

Slipher succeeded in producing spectrograms — pictures of the spectrum. Those pictures usually exhibit features, called spectral lines, that are produced by emission or absorption of light by certain atoms or molecules. These "fingerprints" are what make spectrograms useful to astronomers and physicists. With his shiny new spectrograph, Slipher launched studies of several of the planets. In one instance he found that Venus rotated much slower than previously thought — rather than a period of twenty-four hours, it has turned out to be nearly 225 days.

Although Percival Lowell's top priority was planetary work, in 1909 he asked Slipher to begin a study of the celestial phenomenon known as spiral nebulae (then called only "white" nebulae). A great debate had been swirling among astronomers for more than a century about the true nature of these faint but numerous patches of light. Were they clouds of gas? New solar systems being born in our Milky Way? Or

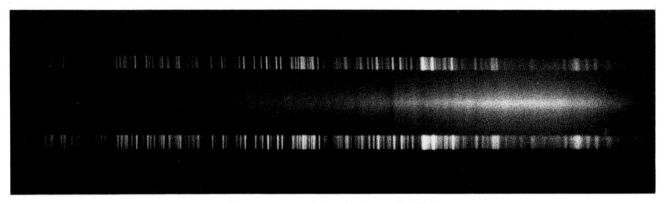

Spectrogram of the Andromeda Galaxy

were they "island universes," that is galaxies entirely separate from our own?

Slipher began the observations but feared that the twenty-four-inch Clark was not strong enough to detect the extremely dim nebulae. A faster lens on the spectrograph camera attached to the scope proved a great boon. Though total exposure times over several nights were still commonly twenty hours or more, he was getting far more detail in his photographic plates. Through the cold nights of the fall and winter of 1912 Slipher shivered in the dome at 7,200 feet elevation, but the rewards were worth the discomfort. With the telescope trained on the Andromeda Nebula, Slipher discovered that the nebula was moving at an incredible speed, three times faster than anything then known.

These results led him on to a spiral nebula in Virgo. When he examined the Virgo plates, Slipher noticed a displacement of lines toward the red end of the spectrum. He knew that if a light source is moving toward an observer, the spectral lines are shifted to the blue. On the other hand, if a light source is moving away, the lines shift to the red. This so-called "red shift" indicated to him that the nebula was receding from the Earth. And it was doing so at an almost unbelievable rate of 620 miles a second, or two million miles an hour, much faster than the velocity he had calculated for the Andromeda Nebula.

In 1914 V.M. Slipher presented his velocities for fifteen spiral nebulae at the American Astronomical Society meeting. He reported that "In the great majority of cases the nebula is receding; the largest velocities are all positive. . . . The striking preponderance of the positive sign indicates a general fleeing from us or the Milky Way." His listeners were so impressed with the speech that they erupted in a standing ovation.

In that "general fleeing" or receding of the spiral nebulae rested the universe-shaking significance of Slipher's discovery. Several years later his velocities were incorporated into equations that showed that the universe was indeed expanding. This assertion, now basic to modern astronomy, holds that when viewed on the scale of the entire universe, galaxies on average are moving away from each other. This idea was nearly as radical as the proposal made centuries earlier by Copernicus that the planets were moving around the Sun rather than the Earth.

A N   E X P A N D I N G   U N I V E R S E

V.M. Slipher with spectrograph on the
Clark telescope

V.M. Slipher's Spectrograph

The spectrograph that V.M. Slipher used to make his monumental discoveries of the speeds of galaxies has been reborn. Lowell Observatory's Paul Roques brought the historic instrument out of mothballs and renovated it, polishing the brass and refitting parts. The spectrograph is on display in the Slipher Building's rotunda.

Made by John A. Brashear of Pittsburgh's Allegheny Observatory, the spectrograph was purchased by Percival Lowell in 1900. It used prisms to separate light into a spectrum. With a camera attached, the spectrograph produced photographs of the spectra. At the time, spectrographs represented a major advancement in astronomy.

When he arrived at Lowell Observatory in 1901, V.M. Slipher mounted the spectrograph on the Clark twenty-four-inch refractor on Mars Hill and began making measurements. His first results were unsatisfactory, and it took nearly a year before he had solved most of the problems. Slipher then produced spectrograms of the planets and, by attaching a faster camera lens and switching from three prisms to one, he obtained good spectrograms of a faint spiral nebulae. The shift of the absorption lines toward the red end of the spectrum led Slipher to deduce that those galaxies were moving away from Earth at unheard-of velocities.

Galaxies Messier 51 and NGC 5195

Not until 1917 would Slipher commit himself to the conclusion that the great speeds of the spiral nebulae indicated that they were indeed "island universes" far beyond our own galaxy. (The island universe debate was finally resolved in the early 1930s, when the great distances of spiral nebulae from Earth were directly measured.)

Though his spiral nebulae velocities were contribution enough to science, Slipher also could claim a number of other "firsts." He discovered that the spiral nebulae were rotating, observed the existence of interstellar gas and dust, and planned and oversaw the successful search for the planet Pluto.

Following Percival Lowell's death in 1916, V.M. Slipher became acting director of Lowell Observatory and ten years later its director. He continued his explorations into the heavens while tending to the myriad administrative details involved with running the observatory. Not the least of his tasks, and one taken seriously by this Indiana farmer's son, was the care of the observatory's cow "Venus" and her calf "Satellite." And in 1928, with a road completed up to the San Francisco Peaks, the robust Slipher built an observing station at 11,500 feet, which operated for about a decade.

Slipher's spiral nebulae discoveries earned him great distinction. In 1933 he was awarded the Royal Astronomical Society's Gold Medal and the National Academy of Sciences Henry Draper Medal. In receiving the Draper award, he thanked several people and also mentioned the blessings of good equipment, favorable skies, and the freedom to pursue his own research. With typical humility, Slipher concluded by saying that "Under such conditions, some one else might have accomplished more, but surely no one could find more pleasure in doing it than I."

V.M. Slipher retired from Lowell Observatory in 1954. He spent the rest of his life in Flagstaff, and died in 1969 at the age of ninety-three.

Thirteen-inch A. Lawrence Lowell telescope used in discovery of Pluto

Pluto

CLYDE TOMBAUGH

Clyde Tombaugh was advised that he was wasting his time looking for another planet in the solar system. If one were there, it would have been discovered a long time ago, an elder visiting astronomer told him.

But Tombaugh persisted in the quest for the ninth planet in our solar system and he found it, in 1930, after only about six months of searching.

V.M. Slipher had hired Clyde Tombaugh, an amateur astronomer from Kansas, as an assistant to help search for the so-called Planet X. At home on the farm in Kansas, when chores allowed, Clyde spent his time grinding mirrors for his own telescopes. Slipher was impressed with the twenty-two-year-old Tombaugh, whom he described as "a young man of the self-made variety." For Tombaugh a job at the observatory founded by his boyhood hero, Percival Lowell, was the opportunity of a lifetime.

With money he made in the harvest, he bought a one-way train ticket and arrived in Flagstaff on January 15, 1929. After dinner (which he thought must have been bear meat) and a night's rest, Tombaugh started his new job. His duties included stoking the big coal and log furnace in the basement, leading visitors on afternoon tours, and readying the new thirteen-inch refractor telescope for the planet search. Lawrence Lowell, then president of Harvard University, had given the $10,000 to buy the telescope. His gift covered the cost of having the thirteen-inch lens ground and figured by C.A. Robert Lundin, of the famous firm of Alvan Clark and Sons who had built the observatory's twenty-four-inch and forty-inch telescopes.

Though it took longer and cost more than was expected, the thirteen-inch lens finally arrived on February 11 and proved a real gem. Tombaugh soon started making one-hour exposures on fourteen-by-seventeen-inch glass photographic plates. After ironing out a few technical difficulties, he began to make three

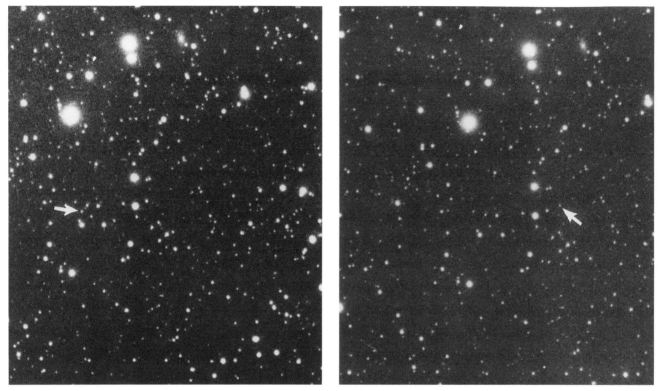

Portions of the plates from which Clyde Tombaugh discovered the planet Pluto. Arrows indicate position of the planet on January 23 (left) and January 29, 1930.

good plates of an area of sky, comparing the best two on a device called a blink comparator. This useful instrument was a microscope on which a pair of plates was mounted, and corresponding areas on each plate then were alternately viewed. As the comparator projected first one plate then the other into the eyepiece, a moving object — a planet for instance — would appear to jump back and forth.

The process may sound deceptively simple, but it was neither simple nor quick. Other objects, such as asteroids, also moved or changed in brightness as the plates were "blinked." And the photographic plates

were literally filled with thousands, sometimes millions, of dots of starlight. "It was as if out of many thousand pins thrown upon the floor one were slightly moved and someone were asked to find which it was," wrote V.M. Slipher and Lowell trustee Roger Lowell Putnam in a 1932 *Scientific Monthly* article.

On the nights of January 23 and 29, 1930, Clyde Tombaugh shot plate numbers 165 and 171 in the constellation Gemini. On February 18 he started to examine them on the blink comparator. At four o'clock in the afternoon his eyes detected an object

Pluto Walk

Mars Hill visitors can take a stroll through the solar system on Lowell Observatory's Pluto Walk, a 350-foot paved path beside the Slipher Building. Signs along the way comprise a scale model of our solar system, illustrating the distances and sizes of the planets in relation to the Sun. Each signpost contains information about that planet and is located to show the planet's average distance from the Sun. Brass markers embedded in the walkway show the maximum amounts the planet deviates from that average as it revolves around the Sun.

The Pluto Walk begins at our Sun, with a diameter one hundred times that of the Earth. The terrestrial planets—Mercury, Venus, Earth, and Mars—stand close together along the path.

Only Earth, ninety-three million miles from the Sun, is exactly situated to allow life as we know it to develop. Had that distance been markedly different, we might not be here.

Beyond Mars lies the main belt of asteroids, strongly influenced by the giant Jupiter. Here the signs become increasingly farther apart as you enter the realm of the outer planets. From Jupiter you journey past Saturn, then to Uranus. Neptune is 2,790 million miles from the Sun, and beyond is the ninth planet, Pluto, discovered at Lowell Observatory in 1930. At the end of the walk stands the dome which houses the Pluto discovery telescope. A bench invites you to relax and contemplate the immensity and diversity of our solar system.

"popping in and out of the background." "That's it!" Tombaugh exclaimed to himself. A check of a third, backup plate confirmed what he had seen. For the next forty-five minutes, Tombaugh said he "was in the most excited state of mind in my life." He let C.O. Lampland know of his discovery and then announced to Dr. Slipher, "I have found your Planet X." That night was cloudy and telescope viewing was poor, so to pass the time Tombaugh went downtown and watched Gary Cooper in *The Virginian.*

Hoping to say confidently that this "very exceptional object" was indeed a planet, Slipher delayed announcement of the discovery to the public. After locating the tiny object in the telescope and tracking and photographing it for three more weeks, the news was finally released on March 13, the

anniversary of Percival Lowell's seventy-fifth birthday. Slipher proudly informed the world that the ninth planet in our solar system had been found at Lowell Observatory, the only planet yet found at an observatory in the United States.

Now that the new planet was verified, what should it be named? Apollo, Atlas, Cronus, Minerva, Perseus, Vulcan, and Zymal were among almost a hundred suggestions submitted. With the public clamoring to be part of the competition, trustee Roger Lowell Putnam urged Slipher and the astronomers in Flagstaff to bestow a name posthaste. Such opportunities at immortality can be delicate matters — Mrs. Lowell asked whether the planet might be called either Percival or Constance (her own name). Slipher realized it was time to act. On May 1 he officially proposed Pluto — the name sent in by an eleven-year-old English schoolgirl and one of Putnam's top choices. Pluto was fitting, Putnam pointed out, because it perpetuated the theme of Roman gods and goddesses for planet names. And the first two letters of the name were conveniently Percival Lowell's initials, a fine memorial. The name was highly appropriate too, for Pluto is the god of the underworld, the domain of the new planet at the outermost reaches of the solar system.

With the thrill of discovery waning, Slipher knew a monumental task still loomed – computing the orbit of the new planet. This challenge bore especially heavily at Lowell Observatory, because of the desire to demonstrate that Pluto was in fact the Planet X that Percival Lowell had predicted twenty-five years earlier. When the orbit was finally computed and then fine-tuned by other observers, Slipher announced with a fair amount of confidence that the planet Tombaugh had found "fits substantially Lowell's predicted longitude, inclination and distance for his Planet X."

Percival Lowell — never one to shy away from the big questions in astronomy — first predicted the existence of an unknown Planet X in a lecture in 1902. His belief in a trans-Neptunian planet was based primarily on the persistence of "residuals," deviations from the expected orbital motion, in Uranus. He speculated that a body of some mass, through its gravitational influence, had to be causing these irregularities in the orbit of Uranus.

Lowell actually conducted several searches for Planet X, albeit quiet and sporadic ones, from 1905 to 1915. The effort involved comparisons of photographs along with complex mathematical calculations to determine a location for Planet X. Though he died without ever seeing the planet, Lowell was close in his predicted location: Pluto was discovered within six degrees of where he said it would be in the sky.

But is Pluto really Planet X, or is another planet still out there awaiting discovery? It is a question being actively pursued by astronomers today. With the discovery of Pluto's moon Charon in 1978 (also in Flagstaff, at the United States Naval Observatory), the mass of Pluto could be calculated. Pluto turned out to be so small that it could not account for the irregular motions observed in the orbits of Uranus and Neptune; thus the search for a tenth planet continues.

After 1930, Clyde Tombaugh scanned the skies from Lowell Observatory for another thirteen years. In the seven thousand hours he spent looking at nearly three-fourths of the sky, he never found another planet. Yet Clyde Tombaugh would have been the last person to discourage would-be planet hunters; he often offered this commandment, based on long experience: "Thou shalt not engage in any dissipation, that thy years may be many, for thou shalt need them to finish the job."

Sun, Mercury, Venus, Earth, Mars, Jupiter, Saturn, Uranus, Neptune, Pluto (not to scale)

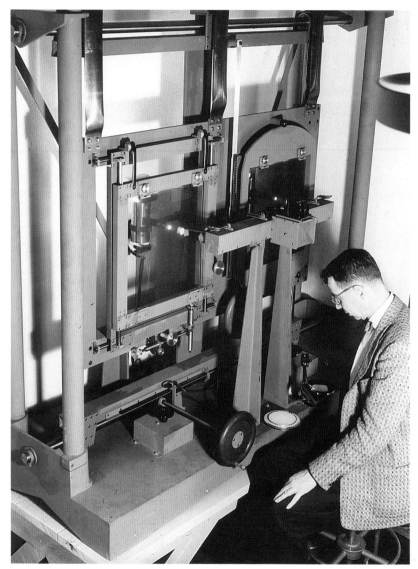

Henry Giclas at the blink comparator at Lowell,
used in his proper motion survey of the stars

A New Era
Begins

After the heady discovery of Pluto, Lowell astronomers had to roll up their sleeves and devote themselves to perfecting calculations of the planet's orbit. Engaged in that effort was Henry Giclas, whose father Eli had helped build the observatory's forty-inch telescope. Henry grew up around Lowell Observatory, and as a college student, during the Great Depression, he landed a job as an observing assistant helping the senior astronomers with Pluto's orbit. The most sophisticated tool they had was a hand-cranked "measuring engine" for taking raw measures of the photographic plates.

Meanwhile, Arthur Adel joined the staff of Lowell, arriving from Michigan in 1933. Adel pioneered infrared spectroscopy, was the first to demonstrate that almost all features in the spectra of the giant planets are due to methane and ammonia, and discovered nitrous oxide and deuterium oxide (heavy-water) vapor in Earth's atmosphere. He also found the "twenty-micron window," a region of increased transparency in the terrestrial atmosphere that lets infrared radiation reach the ground. Through that window, important observations can be made.

After Adel's departure in 1942, exciting discoveries at Lowell Observatory seemed to be growing fainter. But big changes were occurring beyond Mars Hill, and the observatory would soon be swept up in those changes. In the years following World War II, the federal government began funding space and astronomical research in an unprecedented way.

One man in particular—John Scoville Hall—boldly led Lowell Observatory into this new era. A New Englander by birth and education, John Hall seemed to be following in Percival Lowell's footsteps. Hall came to northern Arizona in search of a new home for a forty-inch telescope that belonged to his employer, the U.S. Naval Observatory. When he first arrived in 1952, the road up to Lowell Observatory was still dirt and its small staff of astronomers was working with only three telescopes.

Anderson Mesa is home to Lowell Observatory's modern telescopes and the Navy Prototype Optical Interferometer

Three years later, the Naval Observatory's telescope was relocated to a hill five miles from Lowell Observatory, and Hall came out often to use it.

In 1957, the Soviets successfully launched Sputnik. In quick response the United States government created the National Aeronautics and Space Administration (NASA) in 1958, and the international space race was in full swing. Then-trustee Roger Lowell Putnam recognized John Hall's talents, and in the fall of 1958 Hall was named the fourth director of Lowell Observatory.

He came in quickly, had a vision, and garnered the resources that breathed new life into Lowell's research program. During his nineteen-year tenure, Hall also oversaw significant expansion of the observatory's physical facilities. "More than any single individual," declares Lowell trustee William Putnam, "he molded the Lowell Observatory that enters its second century."

Among his many accomplishments, Hall brought the seventy-two-inch Perkins telescope to Flagstaff and oversaw construction of the forty-two-inch telescope.

Through his supervision, a new instrument shop was built, and in 1960 Lowell Observatory became home for a project to map our Moon under the auspices of the U.S. Air Force. Hall also hired a number of new professional astronomers and introduced an international influence by bringing in a stream of outside, visiting astronomers.

In 1964, a major event took place with the establishment of the Planetary Research Center on Mars Hill, again through funding secured by Hall. At its head was William Baum, formerly of Mount Wilson Observatory. In the ensuing years the center's primary work was the International Planetary Patrol, an ambitious NASA project that involved continuous photographic surveillance of Mars and Jupiter at Lowell and six other locations around the globe. More than a million usable images were produced and stored at the center. Those ground-based patrol photos yielded a great deal of information about fantastic dust storms on Mars during 1971. In addition, outgrowths of the International Planetary Patrol included studies of Jupiter's rotational velocity and Saturn's rings.

As the mountain town of Flagstaff continued to grow, with it came a vexing problem for Lowell astronomers—light pollution. Thus came the next significant development, the dark-sky site on Anderson Mesa. Staff astronomer Harold Johnson had explored several locations outside town, including Padre Butte, A-1 Mountain, and Woody Mountain. After evaluating the ever-important quality of "seeing," along with fire danger, obscuring dust, and the potential for encroaching lights, Johnson recommended Anderson Mesa, about twelve miles southeast of Flagstaff, as the best choice. Hall accepted the recommendation, and by the mid 1960s white domes began to sprout on the mesa top, like mushrooms after a summer rain. Several large telescopes there now permit astronomers to see even deeper into the night sky.

International Planetary Patrol photograph of Jupiter

Meanwhile, Henry Giclas pursued a productive long-term study, a "proper motion" survey of the stars. Assisted by Norman Thomas and Robert Burnham, he set out to duplicate most of Clyde Tombaugh's 1,650 photographic plates and analyze them on a blink comparator, to discern any movement of foreground stars relative to background stars. With the long time lapse of thirty years between Tombaugh's and Giclas's plates, the proper motion of stars could be detected. As a result, thousands upon thousands of stars close to our solar system were identified.

By the time John Hall retired in 1977, Percival Lowell's small observatory in the pines ranked as a full-fledged scientific institution with an international reputation. Astronomers, with ever-more sophisticated tools and technology, found themselves well poised to usher Lowell Observatory into the twenty-first century.

Lowell Observatory's seventy-two-inch Perkins telescope

Those Miraculous Optical Tubes

*I*n 1610 Galileo Galilei took the senators of Venice to the top of a watchtower to demonstrate a new-fangled invention called a telescope. Grasping the utilitarian value of the rudimentary instrument, Galileo wanted to show how easy it was to spy flags flying on ships entering the Venetian harbor.

An outspoken, rabble-rousing, medical school dropout, Galileo built one of his first telescopes by placing two lenses in either end of an organ pipe. When he turned the miraculous optical tube heavenward, he saw some amazing sights—the Moon had mountains and chasms, the giant planet Jupiter was orbited by four satellites, and nearby Venus actually moved around the Sun rather than the Earth. He also marveled at "stars, which escape the unaided sight, so numerous as to be beyond belief."

Astronomers are notorious for their insatiable appetites for light. Any instrument that would satisfy that appetite and that would distinguish far greater detail than could the human eye, was naturally in great demand. From Galileo's time until the present, optical telescopes have remained the major tools of the trade for most astronomers.

The first telescopes were refractors, which use lenses to bring light beams to a focal point. As their size increases, however, refractors become increasingly unfeasible. Sir Isaac Newton subsequently invented an alternative telescope design, the reflector, which does not suffer from these limitations and has other advantages as well. As the name implies, reflectors use mirrors to gather and focus light. By the 1930s a substance known as pyrex was invented and used in telescope mirrors because it was relatively insensitive to temperature changes inside a cold telescope dome. That is a valuable quality because temperature changes can alter the mirror shape, and thus distort the image. The size of telescopes, incidentally, is expressed as the diameter of the lens or mirror.

Although the basic idea behind telescopes has changed little in four centuries, the instruments called detectors that are attached to them have. In the old days, the only detector was the human eye. Then photographic plates were added to gather and record light. After World War II, photoelectric devices were perfected that produced electrical signals in exact proportion to the brightness of the light source being observed. In the 1970s CCDs—charge-coupled devices—combined attributes of both photographs and photoelectric detectors. Charge-coupled devices are semiconductors that record images as an array of typically a million or more small picture elements, or pixels. CCDs are linked to computers that recreate a visible picture on a television monitor.

All this was undreamt in Percival Lowell's time. But from the beginning, he did seek to obtain the best telescopes for his observatory that money could buy. His first major investment was a Clark twenty-four-inch refractor. In 1896, Alvan Graham Clark of the famous Alvan Clark and Sons lens makers, came personally to Flagstaff to install the new lens in the "Clark." Lowell paid $20,000 for the thirty-ton telescope, one of the last refractors made by the firm. It has performed admirably for more than a century, for various research projects and now as the telescope many visitors look through during Lowell Observatory's night sky programs. It is still housed in the handsome ponderosa pine dome built by brothers Stanley and Godfrey Sykes.

The "Clark," now a registered national historic landmark, has been joined by another more modern telescope, the McAllister sixteen-inch. This Cassegrain reflector was installed specifically for public use. The telescope, acquired from Northwestern University, fills in well when the Clark is in heavy demand. Flagstaff philanthropist Frances McAllister donated the funds for the dome in memory of her husband John, an amateur astronomer.

A noteworthy member of the Mars Hill collection is a telescope originally housed there—the Pluto discovery telescope. This is the telescope with which the ninth planet was found in 1930. The thirteen-inch refractor was purchased specifically for the Pluto search by Percival Lowell's brother, Lawrence Lowell. In 1970 the Pluto telescope was moved to Anderson Mesa, but enjoyed a homecoming when it was returned to Mars Hill in 1995 and rededicated. Visitors on guided tours can stroll to the end of the Pluto Walk and enter the basalt-based dome where this historic telescope is housed.

A workhorse of a telescope on Mars Hill is the twenty-one-inch reflector devoted to studies of the luminosity of stars like our Sun. This telescope, which dates to 1953, is housed in a non-traditional "dome" that looks more like a small house with a pitched roof. The roof slides off horizontally, permitting a wide view of the sky. The telescope tube is evidence of a resourceful time when even an old gas line pipe was recycled. Astronomers also occasionally call to duty an eighteen-inch refractor.

The observatory's larger telescopes are located at the dark-sky site on Anderson Mesa. The seventy-two-inch Perkins telescope used to probe the depths of space is the granddaddy of them all. Hiram Mills Perkins, a graduate of Ohio Wesleyan University, joined the staff of his alma mater just as the Civil War broke out. Decreased college enrollments cost him his job, so Hiram returned to the family hog farm. That enterprise, plus Perkins' frugality, led to the accumulation of a tidy fortune. After the war, he rejoined the faculty at Ohio Wesleyan, and by the time of his death in 1924 he had founded Perkins

SOFIA

Ever wondered what to do with a mothballed Boeing 747? Put a telescope in it and fly it high above the Earth, that's what. That's the plan for the Stratospheric Observatory for Infrared Astronomy, or SOFIA, and astronomers from Lowell Observatory hope to be on board.

SOFIA will place a 2.5-meter telescope in a refitted 747. The plane will fly at altitudes up to 45,000 feet, putting the telescope above nearly 99 percent of the water vapor in Earth's atmosphere. The advantage is the ability to see regions of the infrared spectrum that can't be seen by ground-based telescopes.

While most everyone will use SOFIA for infrared spectroscopy, Lowell's Edward Dunham has optical uses in mind. He and James Elliot, of Lowell and the Massachusetts Institute of Technology, are designing and building a high-speed CCD camera for the telescope, to observe occultations of stars by bodies in our solar system. For that use, says Dunham, it is critical to get above the clouds and be in the right place at the right time.

Extensive modifications will be made to the former commercial airplane. An opening will be cut in the rear of the plane through which the telescope will point. The telescope will be fitted with vibration isolators, a gyrostabilizer, and optical tracker. SOFIA is a cooperative project between NASA and the German space agency, DLR.

Observatory and contributed close to $200,000, nearly two-thirds of the cost of a new telescope.

Originally the Perkins was built as a sixty-inch telescope but lacked one important part—a mirror. Clifford Crump, the first director of Perkins Observatory, persuaded the United States Bureau of Standards to undertake an experimental optical program to cast a mirror for it. "Experimental" was the right word. Four mirrors were made and all broke. The fifth, cast in 1928, was a success, but it turned out to be a 3,300-pound, *sixty-nine-inch* disk. Two years later, after the telescope tube was enlarged, the instrument was finally completed and put to work.

The Perkins telescope served well for the next twenty-nine years, but in 1961 in an agreement between Ohio Wesleyan, Ohio State Universities, and Lowell Observatory, it was relocated to Anderson Mesa. In 1965 the sixty-nine-inch mirror was replaced with a seventy-two-inch mirror, and in 1998 Lowell became full owner of the telescope and formed a partnership with Boston University for its joint use and development.

The Perkins has no eyepiece. Instead, astronomers sit before a bank of computers in a warm control room, guiding the telescope's movements and monitoring incoming data gathered on a television-camera system. Hiram Perkins might get a chuckle out of one of the spectrographs attached to his telescope. The big grey metal box bears a sticker that boasts "Don't Laugh, It's Paid For."

Forty-two-inch John S. Hall telescope

Beside the Perkins telescope stands another classic white dome which houses "The John Scoville Hall Telescope." The forty-two-inch reflector, named in honor of the observatory's fourth director who acquired it, is used for stellar research and many other projects.

Anderson Mesa has been a hub of activity in recent years for a major pioneering telescope project—the Navy Prototype Optical Interferometer (NPOI). The telescope consists of mirrors, called siderostats, arrayed in a three-armed Y-shape spread over about fifteen acres of land. This large spread lets NPOI mimic a monstrous telescope 1,475 feet (450 meters) in diameter.

NPOI's siderostats gather light from stars, beam it down vacuum pipes to a collection room where optical carts equalize the paths of light, then recombine the light waves to produce "interference patterns." The recombination is the critical aspect of optical interferometry, because it must be correct to millionths of an inch. NPOI is so exciting because it gives measurements of star positions with unprecedented accuracy, and produces high resolution images that can reveal fundamental stellar properties—the size, shapes, and even changes on the surface of bright stars will be seen. Lowell Observatory is a partner with the U.S. Naval Observatory and the Naval Research Laboratory in NPOI. Lowell's role is site development and operations. Construction began in 1992, "first light" was obtained in 1994, and development continues.

A refurbished 24-inch Schmidt telescope on Anderson Mesa is dedicated to the LONEOS project — Lowell Observatory Near Earth Object Search. The Schmidt, a 1940s-era telescope, was brought from Ohio and refitted with a CCD camera inside the tube. Though not a remarkably large telescope, the Schmidt, with its wide field of view, is well suited to the task of sweeping large portions of the sky each night looking for asteroids or comets that may be headed on a collision course with Earth. Equipped with a powerful CCD camera, the telescope will be able to "see deeper and look at more sky faster," says LONEOS project director Edward Bowell.

Standing apart from the other telescopes on Anderson Mesa is a thirty-one-inch reflector, originally bought by the U.S. Geological Survey for Moon mapping. Lowell Observatory took over the telescope in 1974 and used it intermittently until the main gear cracked. In the late 1980s Lowell engineers completely overhauled the thirty-one-inch: bearings, axles, and the main drive gear were replaced, the mirror was realuminized, and the telescope was cleaned and repainted. With new computerized detectors attached, including a CCD, it

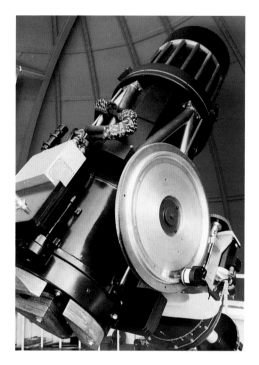

The LONEOS Schmidt telescope

can perform the work of a much bigger telescope. It is used by Lowell researchers and the National Undergraduate Research Observatory. The undergraduate observatory is a consortium of Lowell, Northern Arizona University, and several other colleges whose students gain on-the-job training with a small, research-quality telescope.

Next-generation ground-based telescopes, in which Lowell Observatory is keenly interested, will take advantage of vast improvements in telescope design, in mechanical frameworks and housings, and "active" optics that adjust rapidly to the effects of turbulence. Still, these miraculous optical tubes are only tools for the astronomer. At its heart, astronomy consists of ideas and questions, the sight of a cosmic frontier and the urge to cross it.

NASA Galileo spacecraft image of Io, the most volcanic body in our solar system

Worlds Without End

*P*ercival Lowell was first and foremost a planet man. That tradition of planetary work remains strong at his observatory, but using the latest improvements in technology, astronomers now cast their eyes far beyond our solar system, to stars and galaxies and other heavenly objects and celestial events.

Jupiter and its large satellites have captured the attention of astronomers, especially the innermost "moon" named Io. From the Voyager and Galileo missions they learned, much to their surprise, that Io is a hot, explosive place. That heat, says Lowell astronomer John Spencer, probably derives from the distortion of Io by Jupiter's gravity. Io bends and stretches, creating heat like a twisted coat hanger.

With the Hubble Space Telescope, Spencer and a team of astronomers have gotten spectacular color photographs of Io. Dark colors indicate lava flows from volcanic eruptions, and lighter areas are likely sulfur snow. Along with observations from telescopes on Anderson Mesa and in Hawaii, Spencer has found that Io's volcanoes are big and numerous. Plumes of debris from eruptions are visible shooting off the surface of Io. Spencer has also been studying two other satellites of Jupiter, Europa and Ganymede, which in stark contrast to Io appear to be icy places.

The planet Pluto has naturally remained a prime focus of research at Lowell. Staff astronomer Marc Buie continues to plumb what he calls "the great mystery" of the ninth planet in our solar system. Pluto and its satellite Charon provided information from which Buie created the first map of Pluto's surface, while direct images from the Hubble Space Telescope's faint object camera provided the basis for his second map. Because Pluto is so dim and far away, the maps still show little detail beyond bright spots and dark spots. The bright spots are probably nitrogen frost, with some methane. The dark

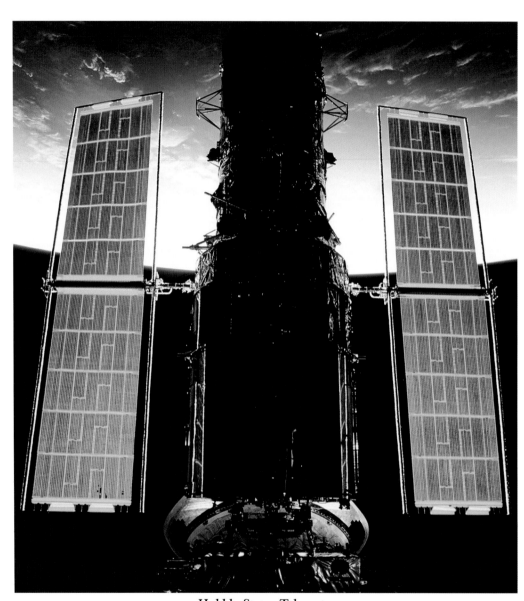

Hubble Space Telescope

WORLDS WITHOUT END

regions, says Buie, are probably "leftovers" of the destruction of methane.

Still, fundamental questions need to be answered: Where did Pluto come from? And is the surface static or changing? Astronomers are using the Perkins seventy-two-inch and the forty-two-inch telescopes on Anderson Mesa for Pluto investigations. In addition, Buie is helping plan a NASA spacecraft flyby mission of Pluto, the only planet that has not yet been visited by a spacecraft. Once launched, the spacecraft will take eight or ten years to reach Pluto. With an imager on board it could supply the first color photos of Pluto, and an infrared spectrometer could reveal a great deal about its chemistry.

An extension of the Pluto work concerns what is known as the Kuiper Belt, a region of very faint objects beyond Neptune. In 1992, researchers started to find objects in the Kuiper Belt. Lowell astronomers have added several dozen more to the list, but there could be at least 100,000 objects about sixty miles or more in diameter, according to Lowell Observatory director Robert Millis.

"We thought we'd completed a reconnaissance of the solar system, but we're not done at all," says Millis, who is fully engaged in the Kuiper Belt search. There may be a vast population of objects out there, including possibly a new class of planets. It is "true exploration," he adds, but because the objects are so faint and so far away the search presents great observational challenges. There are limits to what the largest existing ground-based telescopes can reveal. To discover the true physical properties of the objects, the Pluto mission may also need to fly through the Kuiper Belt region for a close-in look.

While direct views of any celestial object are desirable, Lowell astronomers continue to employ another, indirect method—occultations. When a body in our solar system such as a planet or asteroid comes between the Earth and a star, an occultation occurs.

As the event happens, a shadow, whose cross-section is the size and shape of the occulting object, sweeps across the globe. When the opportunity arises to observe a good occultation, Lowell astronomers hurry to the location, even to high mountains and remote islands, to chase these shadows. Timing is critical, and an occultation may yield only a few seconds of data. But big discoveries can result. From occultations, astronomers, including ones from Lowell, have found that Pluto has an atmosphere and Uranus has a ring system.

Another flourishing research area concerns asteroids. Lowell astronomer Edward Bowell is a preeminent discoverer and namer of asteroids, but as he freely admits that's the easy part. Once discovered they must be followed. Bowell concentrates on calculating asteroid orbits so they can be relocated later. Percival Lowell, incidentally, discovered one asteroid during his career at Lowell Observatory: number 793, which he named "Arizona."

The asteroid work has taken on a most practical application—finding ones that might be on a collision course with Earth. This is the goal of the Lowell Observatory Near-Earth Object Search, or LONEOS.

Consider the possibility. In seconds, a rock some six miles in diameter hurtles to Earth at a thousand times the speed of an automobile driving down the highway. It strikes with the explosive force of 100 million megatons of TNT and blasts a monstrous crater, shattering and cooking surrounding bedrock. Dust rises high into the atmosphere and blocks the Sun, first darkening, heating, and acidifying the environment, then rendering it a sunless, cold place. Half of all life forms on the planet die. The latest gloom-and-doom sci-fi movie? No, this is the scenario scientists paint of the impact event that they think caused the extinction of dinosaurs on Earth sixty-five million years ago.

Craters on Earth and our moon serve as tangible evidence that significant impacts occur on average every 200,000 to 300,000 years. "It is certain," says Bowell, "that another one will occur. We just don't know when."

The LONEOS project scans the skies to find those objects coming within a few million miles of Earth. Orbits are then computed to determine if and when they will intersect Earth's orbit. LONEOS looks specifically for asteroids, and comets, a half mile (one kilometer) or more in diameter, of which there are an estimated 1,000 to 2,000.

Using a 24-inch Schmidt telescope equipped with a CCD camera, sections of the sky are photographed on every clear night. Three, one-minute exposures are made of each section, then the computer compares them to detect motion—much like the tedious old method of "blinking" photographic plates on the blink comparator, except it's all done automatically.

Once a moving object is detected, it's the speed that makes it interesting. "Something going fast or something going in the opposite direction [of common main-belt objects] wakes us up," notes Bowell. LONEOS data goes to the Minor Planet Center in Cambridge, Massachusetts, where it can be transmitted worldwide so others can follow up the observation.

By observing the entire dark sky every month, within ten years, says Bowell, "we hope to see 50 to 60 percent of near-Earth asteroids that could be hazardous to us."

In addition to discovering potentially destructive objects, LONEOS is also amassing a vast amount of data on millions of stars, which collaborators find useful.

Despite their dramatic potential, asteroids have always been the wallflowers of celestial society. These "vermin of the sky," as they've been called, are outshone by their flashier cousins, the comets. Comets spend most of their time as frozen rock and ice on the outskirts of the solar system. They catch our attention when they fly into the inner solar system, their bodies streaming gas and dust in their heads (comae) and tails, reflecting light and appearing as glowing wonders. People have vested comets with all sorts of powers, often evil.

We still eagerly anticipate once-in-a-lifetime close encounters, as we did with the dazzling appearance of Comet Hale-Bopp in 1997. Along with Comets Hyakutake in 1996 and Halley in 1986, astronomers have been able to gather a great deal of data. Hale-Bopp, big and bright, provided Lowell astronomers with a long observing campaign that let them make images and measure the brightness. When Hyakutake came closest to the Earth, they trained four telescopes on it for three months. When Halley reappeared, Lowell scientists were the first to determine that the widely accepted rotation period of two days for the comet was incorrect. They announced to the astronomical community that Halley's rotational period was instead seven and a half days. Further studies have revealed that Halley's comet may undergo a very complex rotation in which it simultaneously turns about more than one axis.

This analysis proved "good training" for Hale-Bopp, says comet researcher Dave Schleicher. Its rotation produces spiral features and is more complex than first thought. In composition, however, Hale-Bopp is a "perfectly normal" comet, notes Schleicher.

All this information goes into a long-term study of the properties of comets as a group, a less glamorous but no less important endeavor. After studying more than a hundred comets for more than twenty years, Lowell astronomers summarized results in a major paper published in the journal *Icarus*. Most comets appear to be fairly homogeneous, made of the same chemicals and formed in the same region. But there are rare exceptions,

Comet Hale-Bopp, photographed April 9, 1997

Starlab

Breathless *oohs* and *aahs* fill the dark chamber as constellations magically appear before the wide eyes of the first-graders. Early one Friday morning, in their school library, this class is transported in minutes to the night sky. This sudden transformation is possible through Lowell Observatory's portable planetarium called Starlab.

"It's a space bubble," observes one young man. Starlab does indeed look like a bubble — a big silver dome that can be inflated within a large room anywhere simply by hooking up a fan. On hands and knees, the 20 or so students crawl inside through a tunnel, let their eyes adjust to the darkness, and exclaim as the pinpoints of the Moon, planets, and stars materialize on the ceiling.

"What do you see besides stars?" the guide asks. The students chorus: "Venus, Venus." He points out the waning gibbous Moon, then the three distinguishable stars that form the belt of Orion, the Hunter. Moving across the sky, the group locates Orion's dog, then the Big Dipper, the Little Dipper, the North Star, and more.

They listen attentively to the stories of what the ancients saw in the constellations — of Queen Cassiopeia and her beautiful daughter Andromeda— and a Native American tale of the three stars in the handle of the Big Dipper, birds hunting bear. Through the eyes of first-graders, dragons and all sorts of other creatures take shape in the night. All it takes is a little imagination, and Starlab.

"a few ringers" says Bob Millis, that have come from somewhere else. For each comet, different parts are active at different times, and the active regions are discrete and localized on the nucleus.

Our Sun, the engine that drives every aspect of our solar system, has also been the focus of a long-term research project at Lowell Observatory. With a photometer attached to the twenty-one-inch telescope on Mars Hill, astronomers Wes Lockwood and Brian Skiff have spent nearly two decades comparing our Sun to other Sunlike stars. "We now know," says Lockwood, "that our Sun varies in brightness only a little bit, under one-tenth of a percent. That's not very much." Still, basic questions remain. Does that steadiness make the Sun special, or not? Will that variation continue? And how much does the waxing and waning of sunspots contribute to climate change on Earth? In related work with astronomer Jeff Hall, the Solar-Stellar Spectrographic program includes several hundred stars and applies ground-breaking technology. Fiber optics are attached to the spectrograph on the forty-two-inch telescope, achieving much greater efficiency in gathering light.

Our Sun is only one of at least 100 billion stars in our galaxy, the Milky Way. The Milky Way, in turn, is one of some ten billion galaxies that can be seen with the largest ground-based telescopes. It is little wonder that Lowell Observatory astronomers seek to understand the nature of stars and the fundamental processes of their evolution—their birth, their lives, and their deaths.

Binary stars may shed some light on these questions. These are double stars that orbit around their common center of mass. Using the fine guidance sensors on the Hubble Space Telescope (HST), Lowell astronomer Otto Franz and collaborators look at the least massive and luminous stars, those at the "bottom end" of stellar distribution, near the border between stellar and substellar objects, called brown dwarfs. About a dozen binary stars are under investigation.

HST's guidance sensors, the telescope's pointing system, operate like an interferometer, obtaining interference patterns from light waves from the stars. By resolving a binary star into its two separate components, Franz has gotten the first definitive orbit on a binary star called Wolf 1062. From this information, mass can be determined. This is critical, Franz declares, because mass is the "driving factor" in what it takes to be a star. From this work, astronomers can better refine what is known about that "edge" where a star is a brown dwarf or a "real" star. Meanwhile, other Lowell astronomers are using the Hubble Space Telescope camera to look at galaxies.

In 1994, Lowell Observatory celebrated a century of astronomical research. But at Lowell, research and education happen at the same time and in the same place. The institution maintains a strong commitment to increasing the public's understanding of the universe and to providing ample opportunity to appreciate the beauty of the night sky. With that in sight, Lowell Observatory marked its centennial with an exciting milestone, the opening of the Steele Visitor Center on Mars Hill.

On May 28, 1994, a hundred years to the day since Percival Lowell arrived in Flagstaff, the doors of the 6,500-square-foot visitor center, built with a generous grant from the Steele Foundation in Phoenix, Arizona, were opened to the public. Adding to the celebration was the arrival of Percival Lowell's shiny red 1909 Stevens-Duryea touring car.

Inside the center, visitors participate in hands-on interactive displays of an exhibit called Tools of the Astronomer. Telescopes, photometers, imaging, spectroscopy, amateur and professional astronomy, and even a simulated observatory invite exploration. Through lenses, prisms, cameras, and computers, visitors of all ages use sight, touch, and hearing to learn how astronomers view the heavens.

After that, it's time to take a seat in the lecture hall where lively guides give programs and use props, including members of the audience, to demonstrate basic science concepts with the Cosmic Cart. "Who can describe the Sun to me?" asks the guide. Hands shoot up, usually the school-age children, who always seem to have the answer on the tips of their tongues. The Sun, the Moon, Venus, the Milky Way—any and all are fair game for topics that usually ask the important question "why?"

On clear nights, visitors then bundle up in warm jackets and walk along the path to the historic Clark telescope or the McAllister sixteen-inch reflector. Everyone has a chance to look through the eyepiece at something of interest—the planet Jupiter with its red spot, the glorious rings of Saturn, or a star cluster like M16. Throughout the year, eclipses, comets, meteor showers, and other special astronomical happenings also draw crowds up to Mars Hill.

The public is welcome to Lowell Observatory in the daytime, to stroll among the Gambel oaks and ponderosa pines of the peaceful campus. They may also tour the beautiful old Slipher Building rotunda, built in 1916. On exhibit are the small globes on which Percival Lowell drew the "canals" of Mars, and the restored spectrograph V.M. Slipher used to discover first evidence that the universe is expanding. Just outside the Slipher Building, the Pluto Walk leads to the Pluto discovery telescope.

Education extends beyond the bounds of Mars Hill. Lowell Observatory takes its show on the road with the Starlab Portable Planetarium, an interactive astronomy museum that travels to schools all over the state. In addition, a special outreach program takes staff astronomers to Native American schools in northern Arizona. In recognition of its efforts, Lowell has won the Grand Canyon State Award for outstanding quality of public science programs and an award from the U.S. Space Foundation's Education Partnership.

University undergraduates and interns gain on-the-job experience observing with Lowell telescopes under the guidance of professional staff. The Lowell Fellowship Program is a point of pride for the observatory. Funded by the BF Foundation and Friends of Lowell, it provides postdoctoral fellowships for young, recent PhDs. Recipients establish residence at Lowell and pursue whatever research interests they choose.

For more than a century, Lowell Observatory has nurtured a strong tradition of ground-based optical observations of the solar system. But as Percival Lowell would have wished, his observatory has moved on to the cutting edge of astronomy. As it enters the new millennium, Lowell Observatory maintains a constant search for answers to the mysteries of the universe.

The historic Slipher Building

Center of galaxy NGC 1569 taken by the Hubble Space Telescope

Suggested Readings

THE HISTORIC LIBRARY

Ferris, Timothy. *Coming of Age in the Milky Way.* William Morrow, New York. 1988.

Hartmann, William K. and Chris Impey. *Astronomy: The Cosmic Journey.* Wadsworth Publishing, Belmont, CA. 1994.

Hoyt, William Graves. *Lowell and Mars.* University of Arizona Press, Tucson. 1976.

_____ *Planets X and Pluto.* University of Arizona Press, Tucson. 1980.

Lowell, A. Lawrence. *Biography of Percival Lowell.* Macmillan, New York. 1935.

Putnam, William Lowell et al. *The Explorers of Mars Hill: A Centennial History of Lowell Observatory 1894-1994.* Published for Lowell Observatory by Phoenix Publishing, West Kennebunk, Maine. 1994.

Smith, Robert W. *The Expanding Universe: Astronomy's "Great Debate" 1900-1931.* Cambridge University Press, Cambridge. 1982.

Snow, Theodore P. *The Dynamic Universe. Fourth edition.* West Publishing Company, St. Paul. 1991.

Tombaugh, Clyde W. and Patrick Moore. *Out of the Darkness: The Planet Pluto.* Stackpole Books, Harrisburg, PA and Lutterworth Press, Guildford and London. 1980.

Trefil, James S. *Space, Time, Infinity.* Pantheon Books, New York and Smithsonian Books, Washington, D.C. 1985.

Verschuur, Gerrit L. *Impact! The Threat of Comets & Asteroids.* Oxford University Press, New York. 1996.